PARTNERS

UNDER THE SEA

Lynn M. Stone

Rourke Publishing LLC
Vero Beach, Florida 32964

www.rourkepublishing.com

PHOTO CREDITS:
All Photographs ©Marty Snyderman except pg 15 © Lynn M. Stone

EDITORIAL SERVICES:
Pamela Schroeder

Library of Congress Cataloging-in-Publication Data

Stone, Lynn M.
Partners / Lynn M. Stone.
p. cm. — (Under the sea)
Includes bibliographical references (p.).
ISBN 1-58952-114-5
1. Marine animals—Ecology. 2. Symbiosis. [1. Symbiosis. 2. Marine animals—Ecology. 3. Ecology.] I. Title.

QL122.2 .S7934 2001
591.77—dc21 2001019427

Printed in the USA

TABLE OF CONTENTS

UNDERSEA PARTNERS

You can snowboard or snow ski by yourself. But you need a partner to water ski. Some undersea animals need a partner, too. Their partners aren't for sport or friendship. They are for **survival**.

A good deal for both partners: Gobies nibble parasites off a Nassau grouper.

Partners can be two or more animals of the same kind. Some of the most amazing partners, though, are from different **species**!

When animals have partners, at least one animal gets something good. That's called a **benefit**.

Barberfish clean a scalloped hammerhead shark.

HURTFUL PARTNERS

One kind of hurtful partner is a **parasite**. A parasite lives off another animal called the **host**. A parasite usually gets food from the skin, muscle, or blood of the host.

Not such a good deal: A harmful, biting isopod clings to the head of a soldierfish.

Some kinds of barnacles are common **marine** parasites. Barnacles are little animals with hard shells. They attach themselves to many different kinds of marine animals. Barnacles that are parasites often live on corals, crabs, and sea urchins.

Parasites hurt their hosts. Their actions can even kill their hosts.

Most parasites live inside their hosts, but these parasite scallops are covering the red crab.

A trumpetfish hangs with a favorite partner, a tiger grouper.

Remoras hitch a ride on a blue shark.

HARMLESS PARTNERS

Some barnacles live on the flippers of whales and sea turtles. The barnacles don't hurt the big animals. The animals' skin or shell gives the barnacles a home. That's a benefit for the barnacles.

Remoras are small fish. With their sucking mouths, they hold onto sharks. They don't hurt the sharks. When sharks feed, remoras let go. Then they feed on leftovers from the sharks' meals.

Barnacles live harmlessly on the shell of this nesting loggerhead turtle that will soon return to the sea.

The anemonefish lives among the stinging **tentacles** of sea anemones. Its partner, the sea anemone, gives the anemonefish safety. Fish that might attack the anemonefish stay away. They could be hurt by the anemone. The anemonefish, though, is not hurt by anemone stings.

Fish like the bluebottle find safety in the same way. They live among the stinging tentacles of jellyfish and their cousins.

The anemonefish lives safely near the magnificent sea anemone's tentacles, which are a threat to most fish.

BENEFITS FOR BOTH

Some marine partners offer benefits for each other. Large fish, for example, know the wrasse. The wrasse is a small cleaning fish. Wrasses grab and eat parasites from the skin of larger fish. The larger fish gets rid of harmful parasites. The wrasse gains a snack. Some fish often swim to cleaning "stations." These are places where cleaning fish gather with fish that want to be cleaned.

A bluestreak cleaner wrasse cleans the mouth of a coral grouper.

Fish aren't the only cleaners. Cleaning shrimp clean up the mouths of certain fish.

The red snapping shrimp lives under the stinging tentacles of ringed anemone. These partners offer benefits for each other. The anemone's tentacles protect the shrimp. The shrimp protects the anemone from little predators such as fireworms.

Like a dentist, a cleaning shrimp cleans the mouth of a black-spotted toadfish.

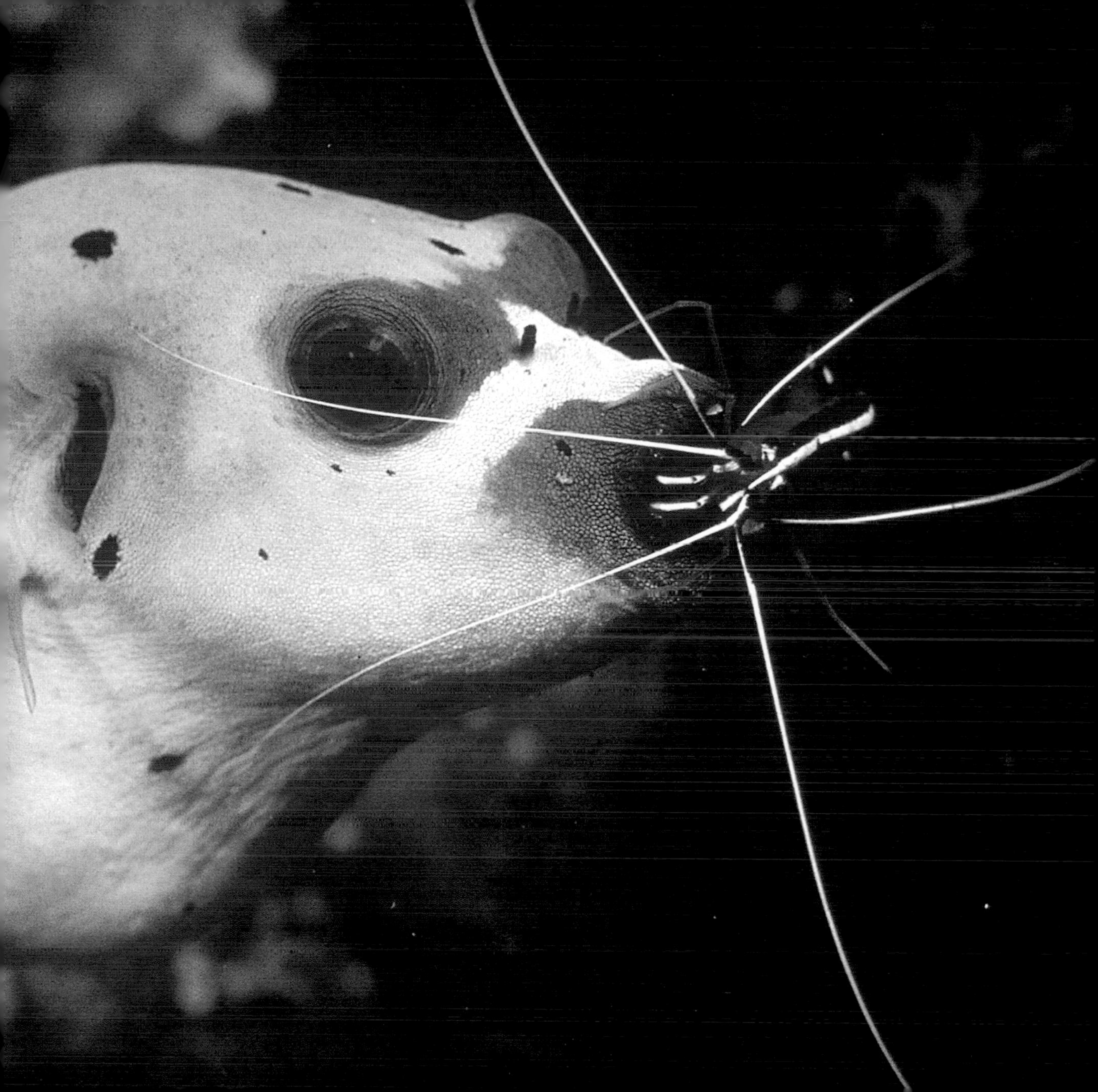

OTHER PARTNERS

The trumpetfish and grouper are strange partners. The trumpetfish seems to hide behind the grouper to hunt small fish.

Humpback whales often partner together for the benefit of the entire group. The huge whales blow bubbles underwater. The bubbles surround schools of fish and frighten them to the ocean surface. There the whales easily catch and eat them.

GLOSSARY

benefit (BEN eh fit) — something of value

host (HOHST) — an animal that is used by another animal for support, protection, or food over a period of time

marine (meh REEN) — of the sea

parasite (PAYR eh syt) — an animal that harms another animal by living on or in it

species (SPEE sheez) — within a group of closely related animals, such as sharks, one certain kind (*reef* shark)

survival (ser VYV el) — living; staying alive

tentacles (TEN te kelz) — a group of long, flexible body parts that usually grow around an animal's mouth and are used for touching, grasping, or stinging

INDEX

Further Reading

Facklam, Margery. *Partners for Life: The Mysteries of Animal Symbiosis, Volume 1*. Little, Brown, 1991
Marquitty, Miranda. *Ocean*. Dorling Kindersley, 1995
Stone, Lynn M. *Fish.* Rourke Publishing, 1993

Websites To Visit

Biology Education at http://www.top20biology.com
Ocean voice at http://www.ovi.ca/

About The Author

Lynn Stone is the author of over 400 children's books. He is a talented natural history photographer as well. Lynn, a former teacher, travels worldwide to photograph wildlife in their natural habitat.